W9-DBK-743

BL 7.4
Pts 1.0

SMART SCIENCE

Electricity and Magnetism

Robert Snedden

Heinemann Library
Chicago, Illinois

© 1999 Reed Educational & Professional Publishing
Published by Heinemann Library,
an imprint of Reed Educational & Professional Publishing,
100 North LaSalle Street, Suite 1010
Chicago, IL 60602
Customer Service 888-454-2279
Visit our website at www.heinemannlibrary.com

Text designed by Visual Image
Cover designed by M2
Illustrations by Paul Bale and Jane Watkins
Printed in Hong Kong/China

03 02 01 00
10 9 8 7 6 5 4 3 2

Library of Congress Cataloging-in-Publication Data

Snedden, Robert.
     Electricity and magnetism / Robert Snedden.
          p. cm. -- (Smart science)
     Includes bibliographical references and index.
     Summary: Discusses various aspects of electricity and magnetism,
including static electricity, electrons, lightning, batteries,
cells, conductors, insulators, circuits, magnets, electromagnets,
electric motors, and more.
     ISBN 1-57572-868-0 (lib. bdg.)
     1. Electricity—Juvenile literature. 2. Magnetism—Juvenile
literature. 3. Electricity—Study and teaching (Elementary)-
–Activity programs—Juvenile literature. 4. Magnetism—Study and
teaching (Elementary)—Activity programs—Juvenile literature.
[1. Electricity. 2. Magnetism.] .I. Title. II. Series.
QC527.2.S64   1999
537—dc21   98-49854
                                                                         CIP
                                                                          AC

Acknowledgments
The publisher would like to thank the following for permission to reproduce photographs:
J. Allan Cash, p. 21; Corbis, pp. 8, 10, 22, 27; Quadrant Picture Library/Auto Images, p. 29; Science Photo Library, p. 13; Vaughan Fleming, pp. 6, 17; Charles D. Winters, p. 7; Peter Menzel, p. 9; Alex Bartel, pp. 19, 23; Martin Bond, p. 25; Andrew Syred, p.28; Tony Stone/David E Myers, p. 4.

Cover photograph reproduced with permission of Science Photo Library, (Peter Menzel)

Note to the Reader
Some words in this book are shown in bold, **like this.** You can find out what they mean by looking in the glossary.

CONTENTS

SMALL BEGINNINGS

Press a button and the television comes on. Flick a switch and the room lights up. Turn a dial and the microwave heats up your lunch. All of these things use electricity. The energy of electricity flows through our lives in numerous ways—but what is it?

What Is Electricity?

Electricity is a form of energy that is caused by the movements of some of the tiny **particles** that make up **atoms**. Atoms are incredibly tiny units of matter. The biggest is just half a billionth of a inch across. They make up everything and everyone around us, including yourself. Each atom is made up of a number of even smaller particles. Normally, there are a number of particles called **protons** at its center or **nucleus**. Protons have a positive **electric charge**. Around the outside of the atom is a cloud of **electrons**. Electrons have a negative electric charge. In an atom, the positive and negative charges cancel each other out, so the whole atom is neutral. It has no charge at all.

Imagine how different this street would be if all the electricity was switched off!

4

Static Charges

Sometimes electrons are attracted from the atoms in one material to those in another. Have you ever noticed that sometimes when you comb your hair, it sticks out from your head? Electrons move from your hair to the comb. The comb, which now has more electrons than before, has a negative electric charge. Your hair, which has lost electrons, has a positive electric charge. Opposite charges are attracted to each other, making your hair stick up toward the comb. The build-up of electric charges as they move from one place to another is what causes **static electricity**. It was the very first form of electricity to be discovered.

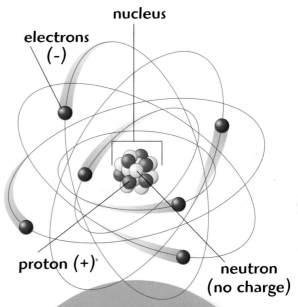

electrons (-)

nucleus

proton (+)

neutron (no charge)

Protons (which have a positive electric charge) and neutrons (which have no charge) form a nucleus at the center of an atom. The nucleus is surrounded by negatively charged electrons.

BRIGHT SPARKS

In Greece around 2,500 years ago, people used polished amber, a fossil formed from pine tree resin, for jewelry and decoration. The Greeks discovered that if amber was rubbed with a cloth, it attracted small pieces of material, such as hair, wool, and feathers.

Two flies were trapped in the sticky resin when this piece of amber became fossilized.

What's in a Name?

We now know that rubbing amber gives it an **electric charge**, in the same way that combing your hair gives it a charge. The Greek word for amber is "elektron," from which we get our words, **electron** and electricity.

For many centuries, **static electricity** was just a curiosity. But then, in the 17th century, investigators such as William Gilbert in England, the first person to use the term "electricity," and Otto von Guerike in Germany began to look into it further. Von Guerike made a large sphere of sulphur that he turned with a crank. He produced sparks by touching the sphere lightly as it spun round.

Moving Charges

Devices were invented to store electric charges. But the only thing scientists did with these was to **discharge** them, by letting the charge flow away. In the 1730s, French scientist Charles du Fay found that he could transfer electric charges from one place to another using a charged-up glass rod. He discovered that sometimes charged objects attracted each other and sometimes they repelled each other. We now know that what happens depends on whether the objects are positively or negatively charged. Opposite charges (a negative and a positive) attract, and like charges (say, a positive and a positive) repel.

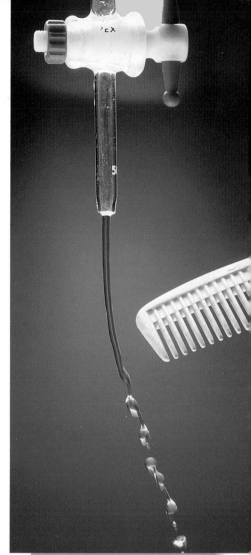

The water is being attracted to the electrically charged comb.

LIGHTNING

Benjamin Franklin, scientist and statesman, realized that the sparks produced in the laboratory were just small-scale versions of the flashes of lightning seen during a thunderstorm. He set out to prove that lightning was electricity, too.

Nice Day to Fly a Kite

In 1752, in a field outside Philadelphia, Pennsylvania, Franklin flew a kite in a thunderstorm. Attached to the kite was a pointed wire, which was tied to a silk thread, with a key on the end. As the storm gathered, Franklin tried something very dangerous. He put his hand near the key and drew electrical sparks from it. (Never try this experiment yourself.) Franklin was very lucky. The next two people who tried it were killed!

Franklin's famous but risky experiment with a kite proved his theory that lightning was a massive **discharge** of electricity.

Franklin's experiment led him to invent the **lightning rod** or **conductor**. Lightning strikes the rod and passes harmlessly through a wire into the ground instead of causing damage to a building.

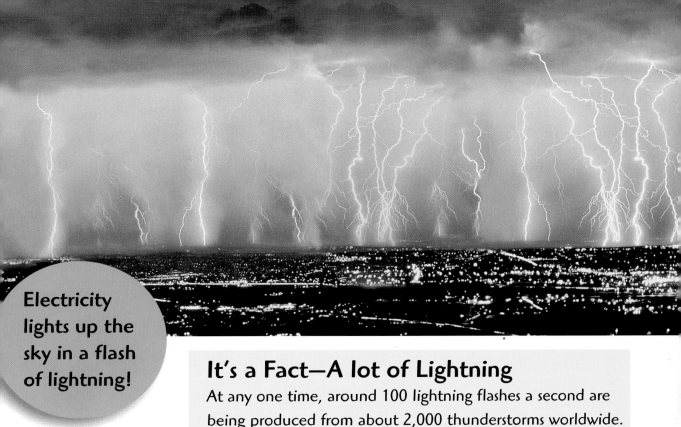

Electricity lights up the sky in a flash of lightning!

Electric Explosion

Lightning occurs when ice **particles** swirling around high inside a cloud become charged with **static electricity**. Air is very resistant to electricity, and it takes a lot of energy to send an electric charge through the air. The negative charge that builds up on a cloud during a thunderstorm can reach millions of **volts**. This is enough to strip **electrons** from the **atoms** in the air, making a passage along which a lightning bolt can travel. Lightning sends an explosion of electrical energy between the cloud and the ground.

Try This—How Far Away Is That Flash?
You need: a thunderstorm and somewhere warm and dry to watch it from!
What to do: The discharge of electricity that produces the lightning also makes thunder as the charge travels through the air. Time how long it takes to hear the thunder after seeing the flash of lightning. Sound travels about 1 mile (1.6 kilometers) in 5 seconds. How far off is the storm?

A SOURCE OF POWER

Electricity is not much use unless it can be put to work. The first thing that is needed is a steady, reliable supply of electrical energy. One such source is the **battery**.

Cells and Batteries

The first battery was invented in the 18th century by the Italian scientist Alessandro Volta. Volta was investigating claims by another Italian, Luigi Galvani. Galvani said that he had discovered that animal muscles generated electricity.

Volta discovered that, purely by accident, Galvani had made an **electric cell**. An electric cell has three parts—a negative **electrode**, a positive electrode, and an **electrolyte**. The electrolyte is a liquid or paste through which electricity can travel. **Electrons** are released by the negative electrode and taken up by the positive electrode. In Galvani's case, the metal plates he used were the electrodes, and the wet muscle was the electrolyte. The source of electricity was the metal plates, not the muscles.

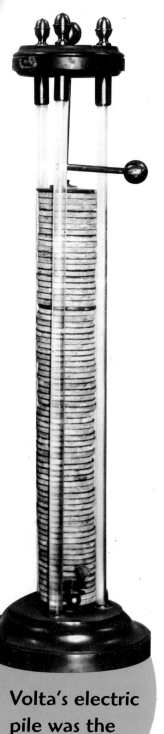

Volta's electric pile was the first battery.

Volta built a pile of cells made up of copper, saltwater-soaked pasteboard, and zinc disks, one after the other in that order. The zinc is the negative electrode, the copper the positive, and the pasteboard the electrolyte. When the pile of cells were connected, they provided a steady flow of electricity. Volta's pile was the first battery—a series of cells linked together.

Dry Cell and Wet Battery

What most people refer to as a battery—the kind that powers a flashlight or stereo cassette player, for example—is actually a type of cell. This is the dry cell, invented by French engineer Georges Leclanché in 1865. The negative electrode is the outer case of zinc, with a central rod of carbon acting as the positive electrode. A paste of ammonium chloride inside the case is the electrolyte.

A car battery is a true battery because it is made up of a number of cells. The cells are made of lead and lead oxide plates bathed in an electrolyte of sulfuric acid and linked together. The chemical reaction that produces the **current** converts the plates to lead sulfate. Running the car engine powers a **generator** that feeds current back to the battery, reversing the process so the battery can be used again.

Electricity is made when the plates and the sulfuric acid in a car battery react together.

It's a Fact— Battery Boom!

Volta made his first battery in 1800. By 1900, producers in the United States were turning out more than 2 million batteries every year.

Try This—The Electric Lemon

You need: a lemon, a piece of zinc, copper wire, a flashlight bulb, and two lengths of **insulated** wire

What to do: Push the zinc and copper wire part way into the lemon. Connect the insulated wires to the metals sticking out of the lemon. Then connect them to the flashlight bulb. The bulb should light up. Congratulations—you've made an electric cell!

ELECTRICITY ON THE MOVE

The electricity we use every day flows in a steady stream rather than in sudden sparks. This is called **current** electricity.

Conductors and Insulators

A current of electricity is simply a flow of charged **particles**, such as **electrons**. We can think of a **battery** as a kind of pump, giving a push to the electrons to get them flowing. This push is measured in **volts**. The higher the **voltage**, the bigger the push the electrons get, and the stronger the current.

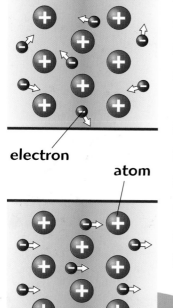

electron

atom

Electrons can flow through a metal easily, moving from one **atom** to another. This is what makes metals such good **conductors** of electricity. The electrons in an electric current drift through the metal, stopping and starting, rather than rushing through. Think of them as being like a row of marbles or ball bearings. If you nudge the marble at one end, the marble at the other end moves at once, even though each individual marble in the row hardly moves at all. In the same way, an electric signal moves very rapidly along a conductor, like a message being passed from electron to electron.

When there is no electric current, electrons move randomly between atoms. When the current is switched on, the electrons drift in the same direction.

However, the electrons do not flow completely freely. Before an electron can move very far, it will probably collide with one of the atoms of the metal. This slows the electron down and may even change its direction.

The pole-type **transformers** at this electricity power station have ceramic disks and concrete pedestals as insulators.

It's a Fact—Superconductors

When many metals and some metal-containing **compounds** are cooled to very low temperatures, around -454°F (-270°C), their resistance drops off to practically zero. This is called superconductivity. The search is on for materials that will be superconducting at everyday temperatures. Scientists have already discovered superconductors that will work at -153°F (-103°C). Superconducting computers would be much faster than anything we use today because there would be no resistance to slow down the flow of information through their electronic **circuits**.

When an electron collides with an atom, some of the electron's energy will be converted to heat. This loss of electrical energy is called **resistance**. The filament in a light bulb resists the flow of electricity through it, and the heat produced makes it glow. Materials with a high resistance, such as plastics, will not let the electrons flow through at all. A material that blocks the passage of electricity is called an **insulator**. Electrons in an insulator are bound tightly to the atoms and cannot easily be started moving. Porcelain is a good insulator and is often used to shield electrical cables.

Try This—To Flow or Not to Flow

You need: your electric lemon (see p. 11), a piece of insulated wire, a variety of objects, such as a paperclip, a plastic spoon, and a fork

What to do: Disconnect one of the insulated wires from the lemon, and attach the new wire. Place various objects between the wires to complete the circuit. Which objects allow the current to flow to the light bulb?

CIRCUITS

Electrical energy needs a channel to flow through before it can be used to produce different types of energy, such as light, heat, and sound. This channel is called a **circuit**.

Series and Parallel Circuits

A simple electric circuit usually has four things: 1) a source of electrical energy, such as a **battery**; 2) something to make use of that energy, such as a light bulb; 3) **conductors**, such as metal wires, to carry the energy from the power source to where it will be used; 4) a switch or some other means of controlling the flow of energy.

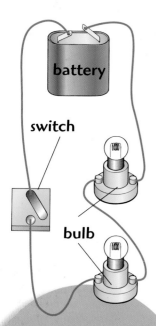

In a series circuit, components are connected end to end. If there is a break anywhere the whole circuit fails.

Anything that conducts electricity can be connected into a circuit, such as a light bulb, a microwave, or a computer. When you put new batteries in a flashlight and turn it on, you are completing a circuit. The flashlight is a simple example of a series circuit. All of the parts of the circuit are connected one after the other, and there is only one path the **current** can take. If one part of the circuit fails, the flow of the current is interrupted, and nothing else in the circuit will work.

In a parallel circuit, components are connected side by side and share the current from the battery. If there is a break in one section, the rest of the circuit will still work.

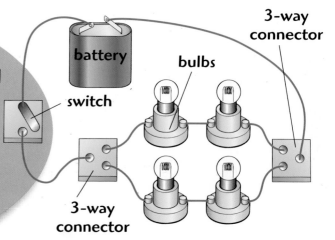

It's a Fact—Ample Amps!

The current flowing in a circuit is measured in **amps**, named after André Ampère, a French scientist. One amp is equal to about 6 billion billion **electrons** per second!

In a parallel circuit, the current may run along two or more paths that are connected side by side. Each path in the parallel circuit is separate.

The circuits in your home are wired in parallel. This means every electrical appliance will not shut down when one light bulb blows!

Circuit Breakers and Fuses

Circuit breakers and **fuses** are devices inserted into a circuit to protect it from a sudden increase in electrical energy. A circuit breaker is like a switch that is triggered when excess current is detected, cutting off the supply of electricity. When the fault has been repaired, the switch is reset. A fuse is a weak link in the circuit. It is designed to melt if too high a current passes through it. This breaks the circuit and stops the current from flowing.

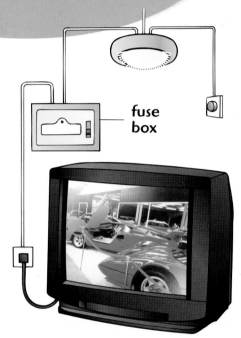

fuse box

Try This—Make a Switch

You need: insulated wire, a paperclip, a paper fastener, a piece of cardboard, a flashlight bulb, and a **battery**

What to do: Push the paper fastener through the cardboard and connect it to the bulb using a length of insulated wire. Connect the bulb to one battery terminal in the same way. Connect the paperclip to the other battery terminal. Slide the paperclip onto the cardboard so it can be swiveled to make contact with the fastener. Switch off the light by sliding the paperclip away from the fastener.

MAGNETISM AND ELECTRICITY

In 1820, the Danish scientist Hans Christian Øersted noticed that the needle on a compass moved when it was brought near a wire carrying an electric **current**. It was soon discovered that electric currents could produce magnetic effects.

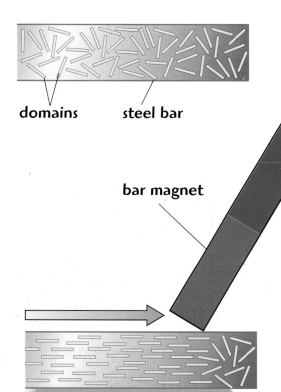

domains   steel bar

bar magnet

In unmagnetized steel, the magnetic regions are jumbled. The bar magnet pulls these domains in the same direction, making the steel magnetic.

What Is a Magnet?

A **magnet** is something that can attract certain metals, such as iron and nickel. The magnet itself will contain one or the other of these metals. A magnet's forces seem to come from two points, one at each end. These are called the north and south **poles**. When the ends of two magnets are brought close to each other, poles of the same kind (say, south and south) repel, but poles of opposite kinds (north and south) attract.

Lodestone, which contains iron, is a naturally magnetic rock. It was used in the first compasses, which may have been in use in China as early as the 5th century B.C. The word magnetism comes from Magnesia, a region in Asia Minor where lodestone was found.

Mini-Magnets

Scientists believe that inside a magnet, there are a great many tiny groups of **atoms** called **domains** that act like mini-magnets. In an ordinary piece of iron, the domains are all jumbled, and their magnetic effects are canceled out. If you stroke the iron with a bar magnet, the domains line up so they all point the same way, and the iron becomes magnetized. If you hit a magnet hard with a hammer, or heat it up, the domains become jumbled again and the magnetic effect is lost.

It's a Fact—Breaking Magnets

If you break a magnet, each new piece has its own north and south poles. No matter how many pieces you break a magnet into, you can never have a single pole on its own.

Try This—Lines of Force

You need: a bar magnet, a sheet of paper, and some iron filings

What to do: A magnet is surrounded by an invisible force field called the **magnetic field**. We can make these lines of force visible. Put the bar magnet on the paper, and carefully scatter some iron filings around it. The filings are magnetized by the invisible magnetic field, and they line up along the lines of force. Most of the filings will cluster around the poles of the magnet, where the force is strongest.

The lines of force around a magnet link its north and south poles.

ELECTROMAGNETS

Øersted's discovery of the link between electricity and magnetism soon led to the invention of the **electromagnet**—a **magnet** generated by an electric **current**.

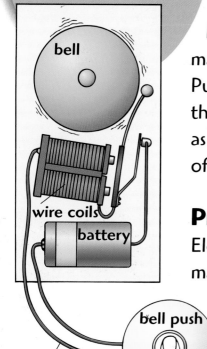

The electric doorbell is just one of the many practical uses of electromagnets.

bell

wire coils

battery

bell push

circuit wire

The Handy Solenoid

The **magnetic field** produced by an ordinary household electric cable is rather weak. A stronger field can be produced by passing an electric current through a coiled insulated wire, called a **solenoid**. The greater the number of turns there are in the coil, the stronger the magnetic field will be. Even a small current can produce a powerful magnet if a large number of turns of wire are used. Putting an iron core inside the solenoid also increases the magnetic effect. The iron is magnetized for as long as the current is flowing. When the current is switched off, the magnetic effect disappears.

Practical Magnetism

Electromagnets are used to lift large masses of magnetic materials, such as scrap iron. They are also used for more delicate tasks in doorbells, loudspeakers, television receivers, and airport metal detectors.

A doorbell makes use of the fact that an electromagnet can be turned on and off. When you ring a doorbell, you complete a **circuit** to turn on an electromagnet. An iron hammer is attracted to the magnet and strikes a bell.

When the hammer moves, it breaks the circuit, and the electromagnet is turned off again. The hammer springs back, and as it does so, it completes the circuit once more. So the magnet is switched back on, attracting the hammer again, and the whole process repeats until you take your finger off the button!

Electromagnets are also very important in electric **motors** and electricity **generators,** as we shall see on the following pages.

Try This—Make an Electromagnet

You need: insulated copper wire, an iron nail, and a **battery**

What to do: Wind the copper wire tightly around the nail, making as many turns as you can. Connect the ends of the wire to the battery. See what you can pick up with your electromagnet.

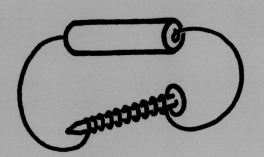

It's a Fact—Flying Trains?

Maglev trains use electromagnets to float above the tracks, giving a smooth, fast, frictionless ride. Speeds of 310 miles (500 kilometers) an hour have been reached in tests carried out in Japan.

Magnetic levitation (Maglev) passenger trains use electromagnets to "fly" above the track.

ELECTRIC MOTORS

Our lives would be very different without electric **motors**. In most homes, electric motors are used every day. Refrigerators, vacuum cleaners, CD players, washing machines, and power tools all have electric motors. They turn electrical energy into movement.

Motors and Magnetism

As we have seen, opposite magnetic **poles** attract each other and like poles repel. Also, if you pass an electric **current** through a wire, you create a **magnetic field**. In the early 19th century, British scientist Michael Faraday was the first to have the idea of putting magnetism and electricity together to produce movement. The first working electric motor was made by U.S. scientist Joseph Henry in 1830.

The magnetic force pushes one side up, the other side down.

Mounting the coil of wire on a shaft allows it to spin around.

Current flows through a coil of wire.

This diagram shows the basic layout of a simple electric motor.

In a simple electric motor, a loop of wire is attached to a metal shaft and is then suspended between the two **poles** of a **magnet**. An electric current is sent through the wire, turning it into an **electromagnet**. The loop twists toward one of the poles of the magnet. When the current is reversed, the magnetism is reversed too, and the coil of wire twists the other way.

Making It Move

A commutator is a simple device that reverses the flow of the current every time the loop makes a half turn. It looks like a copper ring that has been split in two. By switching the direction of the current back and forth in this way, a steady movement of the loop is produced as it rotates on the shaft. This movement can be used to operate a machine. Electric motors generally have many coils of wire and make use of electromagnets rather than ordinary magnets to produce more power.

This streetcar in England picks up its power from the cables running above the track.

It's a Fact—Going Up!

Without the electric motor, the skyscrapers that tower above our streets just would not be possible. Early elevators were steam-powered, but by 1880, the first electrically powered elevators appeared. Today, high-speed elevators can travel at stomach-churning speeds of 20 miles (30 kilometers) an hour or more.

Try This—Wrong-way Motor!

In a motor, magnets move wires. Here, the wires move the magnet, but the principle—using electricity and magnetism to produce movement—is the same.

You need: a compass, a piece of cardboard, a **battery**, copper wire, and two lengths of **insulated** wire

What to do: Mount the compass on the cardboard and wind the copper wire around it. Do about 20 turns—not so much that you hide the compass needle! Connect the ends of the wire to the battery using the insulated wires. The current flowing through the copper wire deflects the compass needle. The stronger the current, the more the needle will move.

GENERATORS

A **generator** is a machine that works in the opposite way that an electric **motor** works. Whereas a motor turns electricity into movement, a generator turns movement into electricity.

Introducing Induction

In 1831, English scientist Michael Faraday and U.S. scientist Joseph Henry, working separately, discovered that if a **conductor** is moved through a **magnetic field,** an electric **current** is produced in the conductor. The magnetic field is said to induce the current in the wire. This is called **electromagnetic induction**. A generator simply keeps the conductor and the magnet moving in relation to each other to produce a continuous flow of electricity. It also provides a way of tapping the electricity produced so it can be used to provide power for electrical equipment.

Dynamos and Alternators

Most generators actually work by spinning **magnets** between coils of wire. It does not matter whether it is the wire that spins or the magnet. The effect is still the same.

Thomas Edison, inventor of the light bulb, was also the builder of the world's first power station.

A **direct current,** or **DC**, generator is sometimes called a **dynamo.**
A dynamo produces a current that always flows in the same direction.
An **alternator** is a generator that produces **alternating current,** or **AC.**
AC current switches direction very rapidly. In North America, the
current alternates, or changes
direction, 60 times per second; in
Europe, it is 50 times per second.

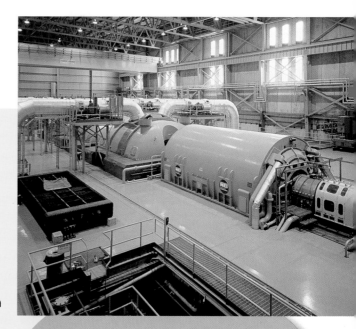

It's a Fact—Power Stations
The first power station, built in Pearl
Street, New York, in 1881, consisted of
three small generators. Each produced
around 10 kilowatts of power, supplying
electricity to 225 houses. A big power
station today can produce a steady
1,300,000 kW of electric power—enough
to run 50 million VCRs!

Here are a huge **turbine**
and generator in a nuclear
power station in Canada.

Try This—Generating Electricity
You need: two bar magnets, two blocks, copper wire, two lengths of **insulated**
wire, and a flashlight bulb

What to do: Set up the magnets so there is a gap between them and space to
move the copper wire between them. You may want
to put them on blocks. Connect the ends of the
wire to the bulb using the insulated wire. Now
move the wire rapidly back and forth between
the magnets. Does the light flicker?

POWER SUPPLY

More than 80 percent of the electricity used in homes and workplaces around the world comes from large **AC generators** driven by steam **turbines**. The turbines are powered by the burning of fossil fuels, by nuclear energy, or by the energy of running water.

Transformers reduce the high voltage of the electricity from power stations to the levels of power needed by factories, railways, shops, farms, offices, and homes.

AC/DC and Transformers

To begin with, all electric power was supplied by **direct current**, or **DC**, in which the **electric charge** flows in one direction only. In 1887, Croatian-American engineer Nikola Tesla devised a generating system that used **alternating current**, or AC. The electrical energy is sent in waves along the power line.

The **voltage** of an alternating current can be made higher or lower by using a device called a **transformer**. "Stepping up" produces a higher voltage, and "stepping down" produces a lower voltage. A transformer can only be used with alternating current, as the current has to be changing for it to work. The beauty of this system is that it means that the voltage can be stepped up to a high level at the power station.

This allows the electrical energy to be transmitted along a greater length of cable than would be possible otherwise. Before the supply reaches the homes and factories where the electricity will be used, another transformer is used to step down the voltage to safer levels.

Power to the People

Electricity is usually generated at some distance from the places where it will actually be used. It must be transmitted along cables, either underground or carried on tall pylons, to where it is needed. High voltages are necessary to provide a sufficent push to get the electrical energy to its destination without too much power loss. The higher the voltage, the lower the **resistance** will be.

Electricity may have to travel a long way from huge power stations like this one to our homes.

Try This—Energy on Tap!

You need: a circle of metal foil, a pencil, a pair of scissors, and a water tap

What to do: Make a hole in the center of the foil for the pencil. Carefully cut some slits, equal distances apart, around the edge of the foil. Bend the pieces to make paddle shapes and put the foil spinner on the pencil. Hold the spinner under a trickling tap. The force of the water will spin your turbine.

ELECTRIC COMMUNICATION

Without electricity, it would be a lot harder for us to keep in touch with people and to learn about events happening around the world. Telephones, televisions, computers, and radios all depend on electricity to make them work.

electromagnet

Electric exchanges

Your telephone receiver turns electrical signals into sounds so that you can hear the voice of your caller.

When you speak into a telephone, your voice is converted into a series of electrical signals that are sent to the person at the other end of the line. Their receiver picks up the electrical signals. This activates an **electromagnet**, which pulls on a thin iron disk. The strength of the signal entering the telephone varies according to the sounds being sent to it. This, in turn, affects the strength of the pull from the electromagnet, which makes the iron disk vibrate. The vibrations pass through the air as the sound of your voice.

iron disk

iron disk

Radio and Television

Radio waves are produced by turning sounds into electric signals and sending them to a **transmitter**. The energy of the signal makes the **electrons** in the transmitter send out pulses of energy that travel in all directions at the speed of light. A radio receiver works in the opposite way. The radio waves strike the **atoms** in the receiver's antenna and produce electric signals. The receiver then translates these signals back into sounds.

Radio waves are also used to transmit television pictures. Again, the radio waves are changed back into electric signals. One of these, the picture signal, is separated into three color signals, which together make a color picture on your screen.

It's a Fact—Homemade Television

Scottish inventor John Logie Baird began his experiments with television in 1923. A cardboard disk made out of a hatbox, a motor mounted on a tea chest, a projection lamp in a cracker can, and lenses from bicycle headlights were all used in his first attempts! In 1925, Baird successfully demonstrated his device in a London department store.

Baird provided the first television programs for the British Broadcasting Corporation in 1929.

Try This—Morse Tapper

You need: a thin, springy strip of metal, a nail, a thumbtack, a piece of wood, a light bulb, a **battery**, and three lengths of insulated wire

What to do: Nail one end of the metal strip to the wood. Bend the other end up slightly and push the thumbtack into the wood underneath it. Attach the nailed down end to the battery using some insulated wire. Use another length of wire to attach the battery to the light. Connect the third wire between the light and the thumbtack. Tap on the metal strip so that it touches the thumbtack. This completes the **circuit** and makes the light flash. Use this to send Morse Code messages!

ELECTRONICS AND THE FUTURE

The design and study of devices, such as computers, that rely on the controlling of the flow of **electrons** is called electronics.

Transistors

In 1947, a team of scientists from the Bell Telephone Laboratories in New Jersey invented the **transistor**. Transistors use materials called **semiconductors**. These come somewhere between **conductors** and **insulators**. Semiconductors can be produced that are either rich or poor in electrons. By putting the two types together, it is possible to make devices that can switch, amplify (increase), or detect electric currents.

Under high magnification we can see the complex circuit on a silicon chip.

Integrated Circuits

Integrated circuits, also called silicon chips or microchips, are used inside computers and electronic games. They are like miniature electronic circuits and are a way of having several components on a single piece of semiconductor. It is possible to have a million transistors or more on a single silicon microchip. Microchips are cheap and reliable and have opened the way to producing small but powerful electronic devices.

It's a Fact—Getting Smaller
The first transistor was about the size of a golf ball. Today, millions can be squeezed into a space scarcely bigger than a postage stamp!

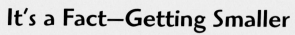

The Electric Future

What does the future hold for electricity? A couple of years ago, the power of electronics was shown when a supercomputer defeated the world chess champion. At least it showed the power of computers to make millions of calculations very fast. However, many computer scientists think it will be another 30 years before we have microchips that can process information as quickly as the human brain.

Is this the car of the future? The EV-1, built by General Motors, is powered entirely by batteries. It's the first electric car to be made available by a major U.S. car manufacturer.

Try This—Pencil Resistor

An important part of any electronics circuit is the **resistor**, which varies the current that flows through the circuit.

You need: a soft graphite pencil lead, a light bulb, a **battery**, and three lengths of **insulated** wire.

What to do: Connect one wire between the battery and the light and the other two from the battery to the pencil lead and from the pencil lead to the light. By moving the wire from one end of the pencil to the other, you can vary the brightness of the light. The longer the length of graphite the current has to flow through, the higher the **resistance**.

GLOSSARY

alternating current (AC) current that switches direction back and forth as it passes through a circuit

alternator type of generator that produces alternating current

amp short for ampere, amps are the amount of electric current flowing through a conductor

atom one of the tiny particles from which all materials are made

battery series of electric cells connected together

circuit arrangement of conducting wires and devices through which an electric current can flow

circuit breaker safety device in a circuit that will cut the supply of electricity if it detects excess current

compound substance that contains atoms of two or more different elements

conductor material through which an electric current can pass easily

current flow of electrically charged particles through a circuit

direct current (DC) current that flows in one direction through a circuit

discharge sudden release of an electric charge

domain small part of a magnetic material that behaves like a tiny magnet

dynamo type of generator, usually one that produces direct current

electric cell something that produces electricity by a chemical reaction between different materials

electric charge property of certain particles that causes them to exert a force on each other. Charge can be either positive or negative. Opposite charges attract and like charges repel.

electrode something from which electric current passes into or out of a conductor

electrolyte liquid or paste through which electrons can flow between electrodes in a cell

electromagnet magnet produced by running an electric current through a coil of wire

electromagnetic induction producing an electric current by moving a conductor through a magnetic field

electron one of the particles that make up atoms; electrons have a negative charge and are the basic particles of electricity

fuse strip of metal or a wire that is designed to melt when an excessive electric current passes through it, breaking the circuit

generator machine that turns mechanical energy into electrical energy

insulator material which blocks the passage of electricity

integrated circuit miniature electronic circuit produced on a single chip of a material such as silicon; also called a silicon chip

lightning rod metal rod that runs from high on a building to the ground to conduct lightning to the earth and prevent damage

magnet any object that forms a magnetic field causing it to attract objects made of iron and some other metals. Magnets have a north pole and a south pole.

magnetic field space around a magnet in which forces of attraction or repulsion act

motor device that changes electrical energy into mechanical energy

nucleus central part of an atom made up of protons and **neutrons**

particles tiny pieces of matter

poles opposite ends of a magnetic **field,** called the north pole and the south pole; like poles repel each other, opposite poles attract each other

proton one of the particles that makes up the nucleus of an atom; protons have a positive electric charge

resistance property of a material that resists the flow of an electric current through it

resistor device that resists an electric current

semiconductor material that has properties somewhere between those of a conductor and an insulator

solenoid electrical conductor consisting of coils through which an electric current is passed to produce a magnetic field

static electricity build-up of electric charge on something

transformer device for increasing or decreasing the voltage of an electric current

transistor device for controlling the flow of electrons through a circuit

transmitter device for broadcasting waves of energy, such as radio waves

turbine engine that is used to drive a generator to produce electricity

volt/voltage measure of the potential energy of an electric current; the "push" given to the electrons to make them flow

MORE BOOKS TO READ

Borton, P. & V. Cave. *Batteries & Magnets.* E D C Publishing. 1995.

Gardner, Robert. *Electricity & Magnetism.* New York: Twenty-First Century Books. 1995.

Gibson, Gary. *Playing with Magnets: With Easy-to-Make Scientific Projects.* Brookfield, CT: Millbrook Press. 1995.

Jennings. Terry. *Electricity & Magnetism.* Chatham, NJ: Raintree Steck-Vaughn. 1995.

Riley, Peter D. *Electricity.* Danbury, CT: Franklin Watts 1998.

Riley, Peter D. *Magnetism.* Danbury, CT: Franklin Watts. 1999.

Wood, Robert W. *Electricity & Magnetism Fundamentals* Broomal, PA: Chelsea House Publishers. 1997.

INDEX